BE BRAVE,
BE CURIOUS,
BE DETERMINED,
OVERCOME
THE ODDS.
IT CAN BE DONE

Will We Survive
on Earth?

STEPHEN HAWKING was a brilliant theoretical physicist and is generally considered to have been one of the world's greatest thinkers. He held the position of Lucasian Professor of Mathematics at the University of Cambridge for thirty years and is the author of *A Brief History of Time*, which was an international bestseller. His other books for the general reader include *A Briefer History of Time*, the essay collection *Black Holes and Baby Universes*, *The Universe in a Nutshell*, *The Grand Design* and *Black Holes: The BBC Reith Lectures*. He died on 14 March 2018.

'Will we survive on Earth?' and 'Should we colonise space?' are essays taken from Stephen Hawking's final book, *Brief Answers to the Big Questions* (John Murray, 2018).

STEPHEN HAWKING

Will We Survive on Earth?

JOHN MURRAY

First published in Great Britain in 2022 by John Murray (Publishers)
An Hachette UK company

1

'Will we survive on Earth?' and 'Should we colonise space?'
are essays taken from *Brief Answers to the Big Questions*,
published by John Murray (2018)

A CIP catalogue record for this title
is available from the British Library

Paperback ISBN 978-1-529-39238-8
eBook ISBN 978-1-529-39239-5

Text design by Craig Burgess

Typeset in Sabon MT by
Palimpsest Book Production Ltd, Falkirk, Stirlingshire

Printed and bound in Great Britain by Clays Ltd, Elcograf S.p.A.

John Murray policy is to use papers that are natural, renewable and
recyclable products and made from wood grown in sustainable forests.
The logging and manufacturing processes are expected to conform
to the environmental regulations of the country of origin.

John Murray (Publishers)
Carmelite House
50 Victoria Embankment
London EC4Y 0DZ

www.johnmurraypress.co.uk

Contents

WILL WE SURVIVE
ON EARTH?

•

In January 2018, the *Bulletin of the Atomic Scientists*, a journal founded by some of the physicists who had worked on the Manhattan Project to produce the first atomic weapons, moved the Doomsday Clock, their measurement of the imminence of catastrophe – military or environmental – facing our planet, forward to two minutes to midnight.

The clock has an interesting history. It was started in 1947, at a time when the atomic age had only just begun. Robert Oppenheimer, the chief scientist for the Manhattan Project, said later of the first explosion of an atomic bomb two years earlier in July 1945, 'We knew the world would not be the same. A few people laughed, a few people cried, most people were silent. I remembered the line from the Hindu scripture, the Bhagavad Gita, "Now, I am become Death, the destroyer of worlds."'

In 1947, the clock was originally set at seven minutes to midnight. It is now closer to Doomsday than at any time since then, save in the early 1950s at the start of the Cold War. The clock and its movements are, of course, entirely symbolic but I feel compelled to point out that such an alarming warning from other scientists, prompted at least in part by the election of Donald Trump, must be taken seriously. Is the clock, and the idea that time is ticking or even running out for the human race, realistic or alarmist? Is its warning timely or time-wasting?

I have a very personal interest in time. Firstly, my bestselling book, and the main reason that I am known beyond the confines of the scientific community, was called *A Brief History of Time*. So some might imagine that I am an expert on time, although of course these days an expert is not necessarily a good thing to be. Secondly, as someone who at the age of twenty-one was told by their doctors that they had only five years to live, and who turned seventy-six in 2018, I am an expert on time in another sense, a much more personal one. I am uncomfortably, acutely aware of the passage of time, and have lived much of my life with a sense that the time that I have been granted is, as they say, borrowed.

It is without doubt the case that our world is more politically unstable than at any time in my memory. Large numbers of people feel left behind both economically and socially. As a result, they are turning to populist – or at least popular – politicians who have limited experience of government and whose ability to take calm decisions in a crisis has yet to be tested. So that would imply that a Doomsday Clock should be moved closer to a critical point, as the prospect of careless or malicious forces precipitating Armageddon grows.

The Earth is under threat from so many areas that it is difficult for me to be positive. The threats are too big and too numerous.

First, the Earth is becoming too small for us. Our physical resources are being drained at an alarming rate. We have presented our planet the disastrous gift of climate change. Rising temperatures, reduction of the polar ice caps, deforestation, over-population, disease, war, famine, lack of water and decimation of animal species; these are all solvable but so far have not been solved.

Global warming is caused by all of us. We want cars, travel and a better standard of living. The trouble is, by the time people realise what is happening, it may be too late. As we stand on the brink of a Second Nuclear Age and a period of unprecedented climate change, scientists have a special responsibility, once again, to inform the public and to advise leaders about the perils that humanity faces. As scientists, we understand the dangers of nuclear weapons, and their devastating effects, and we are learning how human activities and technologies are affecting climate systems in ways that may forever change life on Earth. As citizens of the world, we have a duty to share that knowledge, and to alert the public to the unnecessary risks that we live with every day. We foresee great peril if governments and societies do not take action now, to render nuclear weapons obsolete and to prevent further climate change.

At the same time, many of those same politicians are denying the reality of man-made climate change, or at least the ability of man to reverse it, just at the moment that our world is facing a series of critical environmental crises. The danger is that global warming may become self-sustaining, if it has not done so already. The melting of the Arctic and Antarctic ice caps reduces the fraction of solar energy reflected back into space, and so increases the temperature further. Climate change may kill off the Amazon and other rainforests and so eliminate one of the main ways in which carbon dioxide is removed from the atmosphere. The rise in sea temperature may trigger the release of large quantities of carbon dioxide. Both these phenomena would increase the greenhouse effect, and so exacerbate global warming. Both effects could make our climate like that of Venus: boiling hot and raining sulphuric acid, with a temperature of 250 degrees. Human life would be unsustainable. We need to go beyond the Kyoto Protocol, the international agreement adopted in 1997, and cut carbon emissions now. We have the technology. We just need the political will.

We can be an ignorant, unthinking lot. When we have reached similar crises in our history, there has usually been some-where else to colonise. Columbus did it in 1492 when he discovered the New World. But now there is no new world. No Utopia around the corner. We are running out of space and the only places to go to are other worlds.

The universe is a violent place. Stars engulf planets, supernovae fire lethal rays across space, black holes bump into each other and asteroids hurtle around at hundreds of miles a second. Granted, these phenomena do not make space sound very inviting, but these are the very reasons why we should venture into space instead of staying put. An asteroid collision would be something against which we have no defence. The last big such collision with us was about sixty-six million years ago and that is thought to have killed the dinosaurs, and it will happen again. This is not science fiction; it is guaranteed by the laws of physics and probability.

Nuclear war is still probably the greatest threat to humanity at the present time. It is a danger we have rather forgotten. Russia and the United States are no longer so trigger-happy, but suppose there's an accident, or terrorists get hold of the weapons these countries still have. And the risk increases the more countries obtain nuclear weapons. Even after the end of the Cold War, there are still enough nuclear weapons stockpiled to kill us all, several times over, and new nuclear nations will add to the instability. With time, the nuclear threat may decrease, but other threats will develop, so we must remain on our guard.

One way or another, I regard it as almost inevitable that either a nuclear confrontation or environmental catastrophe will cripple the Earth at some point in the next 1,000 years, which as geological time goes is the mere blink of an eye. By then I hope and believe that our ingenious race will have found a way to slip the surly bonds of Earth and will therefore survive the disaster. The same of course may not be possible for the millions of other species that inhabit the Earth, and that will be on our conscience as a race.

I think we are acting with reckless indifference to our future on planet Earth. At the moment, we have nowhere else to go, but in the long run the human race shouldn't have all its eggs in one basket, or on one planet. I just hope we can avoid dropping the basket before we learn how to escape from Earth. But we are, by nature, explorers. Motivated by curiosity. This is a uniquely human quality. It is this driven curiosity that sent explorers to prove the Earth is not flat and it is the same instinct that sends us to the stars at the speed of thought, urging us to go there in reality. And whenever we make a great new leap, such as the Moon landings, we elevate humanity, bring people and nations together, usher in new discoveries and new technologies. To leave Earth demands a concerted global approach – everyone should join in. We need to rekindle the excitement of the early days of space travel in the 1960s. The technology is almost within our grasp. It is time to explore other solar systems. Spreading out may be the only thing that saves us from ourselves. I am convinced that humans need to leave Earth. If we stay, we risk being annihilated.

•

So, beyond my hope for space exploration, what will the future look like and how might science help us?

The popular picture of science in the

future is shown in science fiction series like *Star Trek*. The producers of *Star Trek* even persuaded me to take part, not that it was difficult.

That appearance was great fun, but I mention it to make a serious point. Nearly all the visions of the future that we have been shown from H. G. Wells onwards have been essentially static. They show a society that is in most cases far in advance of ours, in science, in technology and in political organisation. (The last might not be difficult.) In the period between now and then there must have been great changes, with their accompanying tensions and upsets. But, by the time we are shown the future, science, technology and the organisation of society are supposed to have achieved a level of near-perfection.

I question this picture and ask if we will ever reach a final steady state of science and technology. At no time in the 10,000 years or so since the last Ice Age has the human race been in a state of constant knowledge and fixed technology. There have been a few setbacks like what we used to call the Dark Ages after the fall of the Roman Empire. But the world's population, which is a measure of our technological ability to preserve life and feed ourselves, has risen steadily, with a few hiccups like the Black Death. In the last 200 years the growth has at times been exponential – and the world population has jumped from 1 billion to about 7.6 billion. Other measures of technological development in recent times are electricity consumption, or the number of scientific articles. They too show near-exponential growth. Indeed, we now have such heightened expectations that some people feel cheated by politicians and scientists because we have not already achieved the Utopian visions of the future. For example, the film *2001: A Space Odyssey* showed us with a base on the Moon and launching a manned, or should I say personned, flight to Jupiter.

There is no sign that scientific and technological development will dramatically slow down and stop in the near future. Certainly not by the time of *Star Trek*, which is only about 350 years away. But the present rate of growth cannot continue for the next millennium. By the year 2600 the world's population would be standing shoulder to shoulder and the electricity consumption would make the Earth glow red hot. If you stacked the new books being published next to each other, at the present rate of production you would have to move at ninety miles an hour just to keep up with the end of the line. Of course, by 2600 new artistic and scientific work will come in electronic forms rather than as physical books and papers. Nevertheless, if the exponential growth continued, there would be ten papers a second in my kind of theoretical physics, and no time to read them.

Clearly the present exponential growth cannot continue indefinitely. So what will happen? One possibility is that we will wipe ourselves out completely through some disaster such as a nuclear war. Even if we don't destroy ourselves completely there is the possibility that we might descend into a state of brutalism and barbarity, like the opening scene of *The Terminator*.

• • •

How will we develop in science and technology over the next millennium? This is very difficult to answer. But let me stick my neck out and offer my predictions for the future. I will have some chance of being right about the next hundred years, but the rest of the millennium will be wild speculation.

Our modern understanding of science began about the same time as the European settlement of North America, and by the end of the nineteenth century it seemed that we were about to achieve a complete understanding of the universe in terms of what are now known as classical laws. But, as we have seen, in the twentieth century observations began to show that energy came in discrete packets called quanta and a new kind of theory called quantum mechanics was formulated by Max Planck and others. This presented a completely different picture of reality in which things don't have a single unique history, but have every possible history each with its own probability. When one goes down to the individual particles, the possible particle histories have to include paths that travel faster than light and even paths that go back in time. However, these paths that go back in time are not just like angels dancing on a pin. They have real observational consequences. Even what we think of as empty space is full of particles moving in closed loops in space and time. That is, they move forwards in time on one side of the loop and backwards in time on the other side.

The awkward thing is that because there's an infinite number of points in space and time, there's an infinite number of possible closed loops of particles. And an infinite number of closed loops of particles would have an infinite amount of energy and would curl space and time up to a single point. Even science fiction did not think of anything as odd as this. Dealing with this infinite energy requires some really creative accounting, and much of the work in theoretical physics in the last twenty years has been looking for a theory in which the infinite number of closed loops in space and time cancel each other completely. Only then will we be able to unify quantum theory with Einstein's general relativity and achieve a complete theory of the basic laws of the universe.

What are the prospects that we will discover this complete theory in the next millennium? I would say they were very good, but then I'm an optimist. In 1980 I said I thought there was a 50–50 chance that we would discover a complete unified theory in the next twenty years. We have made some remarkable progress in the period since then, but the final theory seems about the same distance away. Will the Holy Grail of physics be always just beyond our reach? I think not.

At the beginning of the twentieth century we understood the workings of nature on the scales of classical physics that is good down to about a hundredth of a millimetre. The work on atomic physics in the first thirty years of the century took our understanding down to lengths of a millionth of a millimetre. Since then, research on nuclear and high-energy physics has taken us to length scales that are smaller by a further factor of a billion. It might seem that we could go on forever discovering structures on smaller and smaller length scales. However, there is a limit to this series as with a series of nested Russian dolls. Eventually one gets down to a smallest doll, which can't be taken apart any more. In physics the smallest doll is called the Planck length and is a millimetre divided by a 100,000 billion billion billion. We are not about to build particle accelerators that can probe to distances that small. They would have to be larger than the solar system and they are not likely to be approved in the present financial climate. However, there are consequences of our theories that can be tested by much more modest machines.

It won't be possible to probe down to the Planck length in the laboratory, though we can study the Big Bang to get observational evidence at higher energies and shorter length scales than we can achieve on Earth. However, to a large extent we shall have to rely on mathematical beauty and consistency to find the ultimate theory of every-thing.

The *Star Trek* vision of the future in which we achieve an advanced but essentially static level may come true in respect of our knowledge of the basic laws that govern the universe. But I don't think we will ever reach a steady state in the uses we make of these laws. The ultimate theory will place no limit on the complexity of systems that we can produce, and it is in this complexity that I think the most important devel-opments of the next millennium will be.

• • •

By far the most complex systems that we have are our own bodies. Life seems to have originated in the primordial oceans that covered the Earth four billion years ago. How this happened we don't know. It may be that random collisions between atoms built up macro-molecules that could reproduce themselves and assemble themselves into more complicated structures. What we do know is that by three and a half billion years ago the highly complicated molecule DNA had emerged. DNA is the basis for all life on Earth. It has a double-helix structure, like a spiral staircase, which was discovered by Francis Crick and James Watson in the Cavendish lab at Cambridge in 1953. The two strands of the double helix are linked by pairs of nucleic acids like the treads in a spiral staircase. There are four kinds of nucleic acids: cytosine, guanine, adenine and thymine. The order in which the different nucleic acids occur along the spiral staircase carries the genetic information that enables the DNA molecule to assemble an organism around it and reproduce itself. As the DNA made copies of itself there would have been occasional errors in the order of the nucleic acids along the spiral. In most cases the mistakes in copying would have made the DNA unable to reproduce itself. Such genetic errors, or mutations as they are called, would die out. But in a few cases the error or mutation would increase the chances of the DNA surviving and reproducing. Thus the information content in the sequence of nucleic acids would gradually evolve and increase in complexity. This natural selection of mutations was first proposed by another Cambridge man, Charles Darwin, in 1858, though he didn't know the mechanism for it.

Because biological evolution is basically a random walk in the space of all genetic possibilities, it has been very slow. The complexity, or number of bits of information that are coded in DNA, is given roughly by the number of nucleic acids in the molecule. Each bit of information can be thought of as the answer to a yes/no question. For the first two billion years or so the rate of increase in complexity must have been of the order of one bit of information every hundred years. The rate of increase of DNA complexity gradually rose to about one bit a year over the last few million years. But now we are at the beginning of a new era in which we will be able to increase the complexity of our DNA without having to wait for the slow process of biological evolution. There has been relatively little change in human DNA in the last 10,000 years. But it is likely that we will be able to redesign it completely in the next thousand. Of course, many people will say that genetic engineering on humans should be banned. But I rather doubt that they will be able to prevent it. Genetic engineering on plants and animals will be allowed for economic reasons, and someone is bound to try it on humans. Unless we have a totalitarian world order, someone will design improved humans somewhere.

Clearly developing improved humans will create great social and political problems with respect to unimproved humans. I'm not advocating human genetic engineering as a good thing, I'm just saying that it is likely to happen in the next millennium, whether we want it or not. This is why I don't believe science fiction like *Star Trek* where people are essentially the same 350 years in the future. I think the human race, and its DNA, will increase its complexity quite rapidly.

In a way, the human race needs to improve its mental and physical qualities if it is to deal with the increasingly complex world around it and meet new challenges like space travel. And it also needs to increase its complexity if biological systems are to keep ahead of electronic ones. At the moment computers have an advantage of speed, but they show no sign of intelligence. This is not surprising because our present computers are less complex than the brain of an earthworm, a species not noted for its intellectual powers. But computers roughly obey a version of Moore's Law, which says that their speed and complexity double every eighteen months. It is one of these exponential growths that clearly cannot continue indefinitely, and indeed it has already begun to slow. However, the rapid pace of improvement will probably continue until computers have a similar complexity to the human brain. Some people say that computers can never show true intelligence, whatever that may be. But it seems to me that if very complicated chemical molecules can operate in humans to make them intelligent then equally complicated electronic circuits can also make computers act in an intelligent way. And if they are intelligent they can presumably design computers that have even greater complexity and intelligence.

This is why I don't believe the science-fiction picture of an advanced but constant future. Instead, I expect complexity to increase at a rapid rate, in both the biological and the electronic spheres. Not much of this will happen in the next hundred years, which is all we can reliably predict. But by the end of the next millennium, if we get there, the change will be fundamental.

Lincoln Steffens once said, 'I have seen the future and it works.' He was actually talking about the Soviet Union, which we now know didn't work very well. Nevertheless, I think the present world order has a future, but it will be very different.

THERE IS NO
BOUNDARY
CONDITION TO
OUR UNIVERSE.
AND THERE
SHOULD BE
NO BOUNDARY
TO HUMAN
ENDEAVOUR

SHOULD WE COLONISE SPACE?

•

WHY should we go into space? What is the justification for spending all that effort and money on getting a few lumps of moon rock? Aren't there better causes here on Earth? The obvious answer is because it's there, all around us. Not to leave planet Earth would be like castaways on a desert island not trying to escape. We need to explore the solar system to find out where humans could live.

In a way, the situation is like that in Europe before 1492. People might well have argued that it was a waste of money to send Columbus on a wild goose chase.

Yet the discovery of the New World made a profound difference to the Old. Just think, we wouldn't have had the Big Mac or KFC. Spreading out into space will have an even greater effect. It will completely change the future of the human race, and maybe determine whether we have any future at all. It won't solve any of our immediate problems on planet Earth, but it will give us a new perspective on them and cause us to look outwards rather than inwards. Hopefully, it would unite us to face the common challenge.

This would be a long-term strategy, and by long term I mean hundreds or even thousands of years. We could have a base on the Moon within thirty years, reach Mars in fifty years and explore the moons of the outer planets in 200 years.

By reach, I mean in spacecraft with humans aboard. We have already driven rovers on Mars and landed a probe on Titan, a moon of Saturn, but if we are considering the future of the human race, we have to go there ourselves.

Going into space won't be cheap, but it would take only a small proportion of world resources. Nasa's budget has remained roughly constant in real terms since the time of the Apollo landings, but it has decreased from 0.3 per cent of US GDP in 1970 to about 0.1 per cent in 2017. Even if we were to increase the international budget twenty times, to make a serious effort to go into space, it would only be a small fraction of world GDP.

There will be those who argue that it would be better to spend our money solving the problems of this planet, like climate change and pollution, rather than wasting it on a possibly fruitless search for a new planet. I'm not denying the importance of fighting climate change and global warming, but we can do that and still spare a quarter of a per cent of world GDP for space. Isn't our future worth a quarter of a per cent?

We thought space was worth a big effort in the 1960s. In 1962, President Kennedy committed the US to landing a man on the Moon by the end of the decade. On 20 July 1969, Buzz Aldrin and Neil Armstrong landed on the surface of the Moon. It changed the future of the human race. I was twenty-seven at the time, a researcher at Cambridge, and I missed it. I was at a meeting on singularities in Liverpool and listening to a lecture by René Thom on catastrophe theory when the landing took place. There was no catch-up TV in those days, and we didn't have a television, but my son aged two described it to me.

The space race helped to create a fascination with science and accelerated our technological progress. Many of today's scientists were inspired to go into science as a result of the Moon landings, with the aim of understanding more about ourselves and our place in the universe. It gave us new perspectives on our world, prompting us to consider the planet as a whole. However, after the last Moon landing in 1972, with no future plans for further manned space flight, public interest in space declined. This went along with a general disenchantment with science in the West, because although it had brought great benefits it had not solved the social problems that increasingly occupied public attention.

A new crewed space flight programme would do a lot to restore public enthusiasm for space and for science generally. Robotic missions are much cheaper and may provide more scientific information, but they don't catch the public imagination in the same way. And they don't spread the human race into space, which I'm arguing should be our long-term strategy. A goal of a base on the Moon by 2050, and of a manned landing on Mars by 2070, would reignite the space programme, and give it a sense of purpose, in the same way that President Kennedy's Moon target did in the 1960s. In late 2017, Elon Musk announced

SpaceX plans for a lunar base and a Mars mission by 2022, and President Trump signed a space policy directive refocusing NASA on exploration and discovery, so perhaps we'll get there even sooner.

A new interest in space would also increase the public standing of science generally. The low esteem in which science and scientists are held is having serious consequences. We live in a society that is increasingly governed by science and technology, yet fewer and fewer young people want to go into science. A new and ambitious space programme would excite the young and stimulate them into entering a wide range of sciences, not just astrophysics and space science.

The same is true for me. I have always dreamed of space flight. But for so many years I thought it was just that, a dream. Confined to Earth and in a wheelchair, how could I experience the majesty of space except through imagination and my work in theoretical physics. I never thought I would have the opportunity to see our beautiful planet from space or gaze out into the infinity beyond. This was the domain of astronauts, the lucky few who get to experience the wonder and thrill of space flight. But I had not factored in the energy and enthusiasm of individuals whose mission it is to take that first step in venturing outside Earth. And in 2007 I was fortunate enough to go on a zero-gravity flight and experience weightlessness for the first time. It only lasted for four minutes, but it was amazing. I could have gone on and on.

I was quoted at the time as saying that I feared the human race is not going to have a future if we don't go into space. I believed it then, and I believe it still. And I hope I demonstrated then that anyone can take part in space travel. I believe it is up to scientists like me, together with innovative commercial entrepreneurs, to do all we can to promote the excitement and wonder of space travel.

But can humans exist for long periods away from the Earth? Our experience with the ISS, the International Space Station, shows that it is possible for human beings to survive for many months away from planet Earth. However, the zero gravity of orbit causes a number of undesirable physiological changes, including a weakening of the bones, as well as creating practical problems with liquids and so on. One would therefore want any long-term base for human beings to be on a planet or moon. By digging into the surface, one would get thermal insulation, and protection from meteors and cosmic rays. The planet or moon could also serve as a source of the raw materials that would be needed if the extra-terrestrial community was to be self-sustaining, independently of Earth.

What are the possible sites of a human colony in the solar system? The most obvious is the Moon. It is close by and relatively easy to reach. We have already landed on it, and driven across it in a buggy. On the other hand, the Moon is small, and without atmosphere, or a magnetic field to deflect the solar-radiation particles, like on Earth. There is no liquid water, although there may be ice in the craters at the north and south poles. A colony on the Moon could use this as a source of oxygen, with power provided by nuclear energy or solar panels. The Moon could be a base for travel to the rest of the solar system.

Mars is the obvious next target. It is half as far again as the Earth from the Sun, and so receives half the warmth. It once had a magnetic field, but it decayed four billion years ago, leaving Mars without protection from solar radiation. This stripped Mars of most of its atmosphere, leaving it with only 1 per cent of the pressure of the Earth's atmosphere. However, the pressure must have been higher in the past, because we see what appear to be run-off channels and dried-up lakes. Liquid water cannot exist on the surface of Mars now. It would vaporise in the near-vacuum. This suggests that Mars had a warm wet period, during which life might have appeared, either spontaneously or through panspermia (that is, brought from somewhere else in the universe). There is no sign of life on Mars now, but if we found evidence that life had once existed it would indicate that the probability of life developing on a suitable planet was fairly high. We must be careful though that we don't confuse the issue by contaminating the planet with life from Earth. Similarly we must be very careful not to bring back any Martian life. We would have no resistance to it, and it might wipe out life on Earth.

Nasa has sent a large number of spacecraft to Mars, starting with Mariner 4 in 1964. It has surveyed the planet with a number of orbiters, the latest being the Mars reconnaissance orbiter. These orbiters have revealed deep gulleys and the highest mountains in the solar system. Nasa has also landed a number of probes on the surface of Mars, most recently the two Mars rovers. These have sent back pictures of a dry desert landscape. Like on the Moon, water and oxygen might be obtainable from polar ice. There has been volcanic activity on Mars. This would have brought minerals and metals to the surface, which a colony could use.

The Moon and Mars are the most suitable sites for space colonies in the solar system. Mercury and Venus are too hot, while Jupiter and Saturn are gas giants with no solid surface. The moons of

Mars are very small and have no advantages over Mars itself. Some of the moons of Jupiter and Saturn might be possible. Europa, a moon of Jupiter, has a frozen ice surface. But there may be liquid water under the surface in which life could have developed. How can we find out? Do we have to land on Europa and drill a hole?

Titan, a moon of Saturn, is larger and more massive than our Moon and has a dense atmosphere. The Cassini–Huygens mission of Nasa and the European Space Agency has landed a probe on Titan which has sent back pictures of the surface. However, it is very cold, being so far from the Sun, and I wouldn't fancy living next to a lake of liquid methane.

But what about boldly going beyond the solar system? Our observations indicate that a significant fraction of stars have planets around them. So far, we can detect only giant planets, like Jupiter and Saturn, but it is reasonable to assume that they will be accompanied by smaller, Earth-like planets. Some of these will lie in the Goldilocks zone, where the distance from the star is in the right range for liquid water to exist on their surface. There are around a thousand stars within thirty light years of Earth. If 1 per cent of these have Earth-sized planets in the Goldilocks zone, we have ten candidate New Worlds.

Take Proxima b for example. This exoplanet, which is the closest to Earth but still four and a half light years away, orbits the star Proxima Centauri within the solar system Alpha Centauri, and recent research indicates that it has some similarities to Earth.

Travelling to these candidate worlds isn't possible perhaps with today's technology, but by using our imagination we can make interstellar travel a long-term aim – in the next 200 to 500 years. The speed at which we can send a rocket is governed by two things, the speed of the exhaust and the fraction of its mass that the rocket loses as it accelerates. The exhaust speed of chemical rockets, like the ones we have used so far, is about three kilometres per second. By jettisoning 30 per cent of their mass, they can achieve a speed of about half a kilometre per second and then slow down again. According to Nasa, it would take as little as 260 days to reach Mars, give or take ten days, with some Nasa scientists predicting as little as 130 days. But it would take three million years to get to the nearest star system. To go faster would require a much higher exhaust speed than chemical rockets can provide, that of light itself. A powerful beam of light from the rear could drive the spaceship forward. Nuclear fusion could provide 1 per cent of the spaceship's mass energy, which would accelerate it to a tenth of the speed of light. Beyond that, we would need either matter antimatter annihilation or some completely new form of energy. In fact, the distance to Alpha Centauri is so great that to reach it in a human lifetime a spacecraft would have to carry fuel with roughly the mass of all the stars in the galaxy. In other words, with current technology interstellar travel is utterly impractical. Alpha Centauri can never become a holiday destination.

We have a chance to change that, thanks to imagination and ingenuity. In 2016 I joined with the entrepreneur Yuri Milner to launch Breakthrough Starshot, a long-term research and development programme aimed at making interstellar travel a reality. If we succeed, we will send a probe to Alpha Centauri within the lifetime of people alive today. But I will return to this shortly.

How do we start this journey? So far, our explorations have been limited to our local cosmic neighbourhood. Forty years on, our most intrepid explorer, Voyager, has just made it to interstellar space. Its speed, eleven miles a second, means it would take about 70,000 years to reach Alpha Centauri. This constellation is 4.37 light years away, twenty-five trillion miles. If there are beings alive on

Alpha Centauri today, they remain blissfully ignorant of the rise of Donald Trump.

It is clear we are entering a new space age. The first private astronauts will be pioneers, and the first flights will be hugely expensive, but over time it is my hope that space flight will become within the reach of far more of the Earth's population. Taking more and more passengers into space will bring new meaning to our place on Earth and to our responsibilities as its stewards, and it will help us to recognise our place and future in the cosmos – which is where I believe our ultimate destiny lies.

Breakthrough Starshot is a real opportunity for man to make early forays into outer space, with a view to probing and weighing the possibilities of colonisation.

It is a proof-of-concept mission and works on three concepts: miniaturised spacecraft, light propulsion and phase-locked lasers. The Star Chip, a fully functional space probe reduced to a few centimetres in size, will be attached to a light sail. Made from metamaterials, the light sail weighs no more than a few grams. It is envisaged that a thousand Star Chips and light sails, the nanocraft, will be sent into orbit. On the ground, an array of lasers at the kilometre scale will combine into a single, very powerful light beam. The beam is fired through the atmosphere, striking the sails in space with tens of gigawatts of power.

The idea behind this innovation is that the nanocraft ride on the light beam much as Einstein dreamed about riding a light beam at the age of sixteen. Not quite to the speed of light, but to a fifth of it, or 100 million miles an hour. Such a system could reach Mars in less than an hour, reach Pluto in days, pass Voyager in under a week and reach Alpha Centauri in just over twenty years. Once there, the nanocraft could image any planets discovered in the system, test for magnetic fields and organic molecules and send the data back to Earth in another laser beam. This tiny signal would be received by the same array of dishes that were used to transit the launch beam, and return is estimated to take about four light years. Importantly, the Star Chips' trajectories may include a fly-by of Proxima b, the Earth-sized planet that is in the habitable zone of its host star, in Alpha Centauri. Only in 2017, Break-through and the European Southern

Observatory joined forces to further a search for habitable planets in Alpha Centauri.

There are secondary targets for Breakthrough Starshot. It would explore the solar system and detect Earth-crossing asteroids. In addition, the German physicist Claudius Gros has proposed that this technology may also be used to establish a biosphere of unicellular microbes on otherwise only transiently habitable exoplanets.

So far, so possible. However, there are major challenges. A laser with a gigawatt of power would provide only a few newtons of thrust. But the nanocraft compensate for this by having a mass of only a few grams. The engineering challenges are immense. The nanocraft must survive extreme acceleration, cold,

vacuum and protons, as well as collisions with junk such a space dust. In
addition, focusing a set of lasers total-
ling 100 gigawatts on the solar sails will
be difficult due to atmospheric turbu-
lence. How do we combine hundreds of
lasers through the motion of the atmosphere, how do we propel the nanocraft
without incinerating them and how do
we aim them in the right direction?
Then we would need to keep the nanocraft functioning for twenty years in the
frozen void, so they can send back
signals across four light years. But these
are engineering problems, and engin-
eers' challenges tend, eventually, to be
solved. As it progresses into a mature
technology, other exciting missions can
be envisaged. Even with less powerful
laser arrays, journey times to other

planets, to the outer solar system or to interstellar space could be vastly reduced.

Of course, this would not be human interstellar travel, even if it could be scaled up to a crewed vessel. It would be unable to stop. But it would be the moment when human culture goes inter-stellar, when we finally reach out into the galaxy. And if Breakthrough Starshot should send back images of a habitable planet orbiting our closest neighbour, it could be of immense importance to the future of humanity.

In conclusion, I return to Einstein. If we find a planet in the Alpha Centauri system, its image, captured by a camera travelling at a fifth of light speed, will be slightly distorted due to the effects of special relativity. It would be the first time a spacecraft has flown fast enough to see such effects. In fact, Einstein's theory is central to the whole mission. Without it we would have neither lasers nor the ability to perform the calculations necessary for guidance, imaging and data transmission over twenty-five trillion miles at a fifth of light speed.

We can see a pathway between that sixteen-year old boy dreaming of riding on a light beam and our own dream, which we are planning to turn into a reality, of riding our own light beam to the stars. We are standing at the threshold of a new era. Human colonisation on other planets is no longer science fiction. It can be science fact. The human race has existed as a separate species for about two million years. Civilisation began about 10,000 years ago, and the rate of development has been steadily increasing. If humanity is to continue for another million years, our future lies in boldly going where no one else has gone before.

I hope for the best. I have to. We have no other option.

THIS IS NOT THE
END OF THE STORY,
BUT JUST THE
BEGINNING OF
WHAT I HOPE WILL
BE BILLIONS OF
YEARS OF LIFE
FLOURISHING
IN THE COSMOS